ABDUL RAHMAN

Making Money as Freelancer

Copyright © Abdul Rahman, 2018

All rights reserved. No part of this publication may be reproduced, stored or transmitted in any form or by any means, electronic, mechanical, photocopying, recording, scanning, or otherwise without written permission from the publisher. It is illegal to copy this book, post it to a website, or distribute it by any other means without permission.

First edition

This book was professionally typeset on Reedsy. Find out more at reedsy.com

"I dedicate this Book to all the Freelancers around the world to benefit from this Book."

By Abdul Rahman
The Author

Contents

What is Freelancing? and How to become a Freelancer?	1
How to Start and Succeed as Freelance Blogger	3
Remote.com : The great Freelance Marketplace	11
Workhoppers.com : The Growing Local Freelance Startup...	14
New Freelance Marketplace :Linkedin Profinder	17
Fiverr.com The Freelance Marketplace for Professional Gigs	20
Workana.com : The New Freelance Startup Marketplace	24
Seven Secrets of becoming a Successful Freelancer	27
Freelancing in 2018: Prospects and Opportunites	33
The Social Media Importance For Freelancers & Entrepreneurs	35
Posting A Project On Freelance Marketplaces	38

1

What is Freelancing? and How to become a Freelancer ?

Freelancing is self employment Job done by Millions of Freelancers world . A Freelancer is a Person who complete the Projects or Tasks working from home . Freelancers have not any long time job commitment. Freelancers are Hired on Hourly or Fixed Price basis . Suppose if you have a project of content writing or Blogging , You can hire a Professional Freelancers from the pool of verified and talented People regardless of country or origin .

With the Advent of Internet , the world has become a global village which has made it possible for Small Business Companies or Individuals to post their Projects on the sites called Freelance Marketplaces such as Freelancer.com, Elance.com , Odesk.com , Bizreef.com or Guru.com to name a few .

The Freelance marketplaces are substitute to Employment Exchange since they provide the Powerful Platform for the Buyers to connect with Professional freelancers available round the clock working on the Award-ed projects. The Freelancing has become a million Dollar Business since it has sustained and bloomed even in Financial recession when companies were downsizing in the Human Resource but the Freelance Marketplaces were flocked with thousands of Buyers and Freelancers doing various tasks.

Freelancing can be done in various categories but the most popular categories include Writing and Translation, Web Design and Development, Programming, Graphic Designing, Logo Designing, APP Development, Business Plan Writing, Startup Consulting, HR Consulting and Data Entry etc. If you want to be a Freelancer then register on Free-lance Marketplace websites in the relevant category of your choice and Start freelancing. If You want to be a Successful Freelancer[1] then read the Blog Post on the Topic.

2

How to Start and Succeed as Freelance Blogger

Content is king and the content creator is the person bewitches the people who read the read the magical words attracting millions of the internet User across the globe on different platforms such as IPads ,

Smartphones , Desktops , laptops and E-Readers . Everyday thousands of Freelance Bloggers rock the world with worth reading Masterpieces related to the Interest of People .People Reach the fresh content through variety of platforms such as Social Media , Networking , RSS Feeds ,Search Engines and Referrals . The Bloggers are the people who share their thoughts in blog posts usually of 250 to 600 Words . But the Marketing Based blogs may post upto 1000 plus Word blog pots or How to based to share the tips and tools of making the most of Internet .

Some say about tech Startups ,Some blow up with Marketing wand

, Some Share the killer gadgets , Some teach you how to make money from your Blog or Website , Some bloggers also cover the Advertising Modes , Some Play with Code Poetry , Some share the creative thoughts and Design Tips , Some Share the Writing Basics means you can search for any topic blog and its 100% sure that you will find many not few. Blogging in the beginning was limited to sharing your personal Experiences and Happening unlike writing Personal Diaries but now a days it has become the Full Scale Business and Blogs Such as JohnChow.com , TechCrunch.com , Mashable.com ,Huffingtonpost.com having earning Millions of Dollars and the list is not limited almost every blogger makes to three figures some may cross four Figures depending upon the niche and quality of content being Produced . Now a days , every business has got a blog to keep the customers regarding the update , improvements and Promotions regarding the Organization.

If you are a Startup Blogger and have been struggling to succeed as Blogger then do not stop , keep on blogging but learn the basics of Blog-ging from the trend setters in blogosphere , read the inspiring stories of the Successful bloggers to learn what model they Applied to become Successful from Startup blogger . The Following points will help you in becoming an established Blogger with high ranking Blog and swarm of traffic flocking your blog to quench their thrust for particular topic .

1. Domain Name :

Domain registration is the First step to set your foot in blogging planet . but wait! what have you decided to blog about ? This question is they to succeed as Topic for blogging matter the most. You can chose any topic that you love to discuss and topic related to your career, hobby or interest. After your topic selection, you must search for the domain related to your blog. Domains such as .com ,.co or

.org are much better than .info etc domains as these top level domains are easily memorable. Almost ,all bloggers aspire to start their startup blog with a .com domain. If the domain is available then Wow! You are lucky to go the Next Step.

2. Platform and Hosting :

There are Several CMS based Blogging platforms such as WordPress, Squarespace, Typepad, Joomla ,Drupal and Blogger etc. But you have to make choice. It is recommended that you should go with WordPress as it is very popular Blogging platform and the platform of choice for the Pro Bloggers all around the world. Now, when you have decided to go with WordPress, You need a professional host to provide unlimited Space and unlimited bandwidth to your blog. There are several WordPress Optimized Hosts but its recommended that you should choose the reliable hosting services such as BlueHost, Ipage.com, name.com , GoDaddy.com or Hosting24.com to power up your Startup blog. After purchase of Host you may login to your host and install the WordPress from Script Installer Such softaculous etc. If you are not good in Computer Skills you may hire the services of professional Developer and Designer to Install the WordPress for you and Power up with custom logo ,theme and Plugins to bang the world. Next Step is the Content creation.

3. Writing your First Blog Post :

Writing your first blog post related to choice of your subject is very difficult since you do not know how to attract the massive audience with Sticky and attractive Heading and Blog Post . Here you feel the dearth of Skills which can easily be filled by following the Successful Bloggers and reading their Tips and practicing the same . You may use the most searchable Keywords bu using Google or other Search Engines . Keywords are those terms by which the Search engines identify the content across globe and display the relevant content as per search query . It is advisable that you should use the catchy headlines to attract massive readers . Instead of writing "My Success Story as Blogger " You can make it more catchy by writing "How I went from Zero to $10000 Dollars per Month blogging about the things I loved " The latter Headline looks more catchy as compared to Former heading .

4. Content Quality and Frequency :

After writing your first blog post ,you must have realized that how easy it was to write about the thing you love and how to use the keywords in the body of the blog post. The Standard and Error Free Content by the Grammar or Typing Point of few attracts more readers and win appreciation or feedback through Commenting System . You may deploy Commenting System such Disqus ,Intense Debate etc . Some Startup blogs engage multiple authors for multi-authored blogs or Network of Blogs run by the Bloggers . If you intend to engage multiple authors you can hire Freelance authors avail-able at freelance marketplaces . Now , the Frequency of Posting new blog posts to your startup blog . Some bloggers post articles on daily basis ,some bi weekly or some busy blogger may post for fun and post to blogs on weekly basis when they are free . but it is recommended by Blogging Experts to submit posts on daily basis which improves its search traffic ,rank and reputation . Some blogs also submit three to 10 Blog posts dai-ly and receive a regular traffic and high rankings . Its all about content ,quality and SEO (Search Engine Optimization)

5. Search Engine Submission:

Search engines help you to in-crease your organic traffic and your visibility . If your blog ranks higher in search engines related to your blog categories then your blog will be get-ting constant search traffic globally and increasing your page views and at the same time your revenue will multiply provided that you maintain content quality ,standard and SEO (Search engine Optimization) . You should use the proper Keywords and it density in the body and headlines to increase page views . Headline with How to...... attract more visitors as compared Static Headlines . You should also use Google Analytics to have insight and real time statistics of your blog . You can use free or paid Searcj Engine Submission for Marketing .

6. Adverting and Marketing :

Advertising makes your blog heard by the readers through variety of channels and blogs using Ad net-works to monetize their inventory . Your Ad for for the Blog Marketing leads to your website after being clicked by the visitors . This will increase the conversions and your blog viewership. Advertising Networks offer variety of Ad Formats and Tools to attract massive traffic through geo targeting through the Internet Protocol (IP) addresses of the computers with which they access Internet globally. You can opt for PPC (Pay Per Click) CPM (Cost per Million) based ads and Banners .It is recommended that for Contextual Ads select PPC mode and for Banners it is advisable to use CPM based Mode. Advertising Network also offer Backlinks, Link Exchange, Content Marketing, in text ads, hover ads, slider ads and Peel away ads. All these modes have one simple target that is to introduce your blog to massive audience and expand your outreach to readership.

7. Socialize :

Social Media is the powerful Medium to market your Blog and the content you create for the readers on your startup blog. If you do not share with people whatever you have done some remark-able will go unheard until you announce it publicly . This is like a person who succeeded in in the experiment for developing vaccine for any disease but did not announce the same achievement to the people to get benefited from the vaccine. Almost all the Pro Bloggers have active So-cial Media Accounts with various Social Media Websites such Facebook

, Twitter, Linkedin, Yammer ,StumbleUpon, Pinterest etc to share the Photos ,blog posts,tips and updates regarding themselves as well as about the Industries they are working . You should install or ask the Web De-veloper to deploy Social Media Sharing Buttons so that readers may share the article with friends. Social Media is the simplest,Free and easiest way to get massive readership and traffic amazingly at no cost.

8. Email Marketing :

Email Marketing is the effective to turn your friends and contacts into subscribers or Customers if you sell something on your blog such as Service , E-books , software etc . Email Marketing Companies such as Aweber , Constant Contact , MailChimp etc are the best to get started with even the feedburner from google may blast the readers with your latest posts and updates through email . You can also hire a permanent Email Marketing Consultant to create campaigns and send the Articles of mass interest to Subscribers on daily , weekly or Monthly basis depending upon the interest of Readers Choice that whether the reader/subscriber has opted for daily , weekly ,or monthly Newsletter or Updates . You need to care about the reader's interest and don't scrap their inbox with worthless content which may force the subscriber to unsubscribe.

9. Revenue Generation :

Revenue is the Soul to survive as Start-up Blogger almost all Startup blogs have certain revenue Models . Some focus on advertising Models such as Google Adsense , Chitika , Media Net , Infolinks , Some may opt for Affiliate Marketing and earn Com-mission ,Some develop E-Books with Catchy Headlines , Some write Sponsored Reviews for the Companies , Earn Money and Support their Blog to survive . Some sell ad space Directly through BuySellAds , Ad-Sella ,Publicity Clerks . It is recommended that First Advertising Model such Google Adsense , Chitika , Media.net , Infolinks is the best option along with Direct Ad selling Marketplaces such BSA and AdSella. Revenue is the Key to sustain and develop your blog into Money Stream . You can also develop How to .... E-books and Sell or offer Free to increase email subscribers .

By above steps,if taken in proper manner , you will soon be counted as Successful Blogger and your blog rank will get improved in sites such Compete.com , Alexa.com and Similarweb.com . These websites judge the traffic and Social Media Trends as well as the Backlinks . Backlinks are considered the Cashlinks as they refer the visitors to your website and increase conversion and increase revenue . Backlinks can created through commenting on the leading Blogs and Forums . The Comment along with your name ,Site name and URL will be Published and you will be getting constant traffic from the backlinks . Attend some Bloggers Meet-Up and share your story as well lean from the Market Leaders to explore the opportunities .

-

3

Remote.com : The great Freelance Marketplace

Freelance Marketplaces are getting a Tremendous response from the Freelancers around the world as the Technology has bridged the gap between the providers and the Employers through the Powerful platforms of the freelance marketplaces.Freelancers throughout the world are searching various projects of their Interest and earning Millions of dollars through Freelance Marketplaces.

There are several Freelance Marketplace available for both Employers and Freelancers but Remote.com is a very good addition to freelance Marketplaces. It has a very powerful platform to connect Freelancers with the Employers searching for Freelancers.

Establishment: Remote.com was established in 2017 to provide services to freelancers by connecting them with the prospective clients to get the projects done with ease.

Membership: The Membership is free for both the Freelancers and Buyers as they get registered either through Facebook or Google Social login. Such a feature removes the hassle of entering long sign up data and you are up and ready just in seconds. For Buyers remote.com offers three job positing plans : They include : Free , Premium and Elite , However , the freelancers are charged flat 10% on their earnings .

Project Categories : The Buyers can post jobs in multiple categories ranging from Web Development and Programming, Translation, writing blogging and Marketing etc where as Freelancers can apply for freelance projects in above categories .

Payments: The Buyers can pay through Credit cards while Freelancers can withdraw money through ACH , Paypal and International Wire transfers. The payments are very secure and easy for the both buyers and freelancers .

Upside : Remote.com is the Great Freelance Marketplace offering a Powerful freelance Marketplace as it is easy to navigate and free to start working as Freelancer or as a buyer .

DownSide: Remote.com is a good site but lacks an attrcative Design and Payments options . It must include Payoneer and Skrill for Global Freelancers to withdraw their earnings easily .

Future Trend analysis: The Way remote.com has been getting popularity amongst the Freelancers, it is estimated that in future it will be very powerful freelance marketplace for both the Buyers and Freelancers and would be counted among the best freelance Marketplaces over the internet.

4

Workhoppers.com : The Growing Local Freelance Startup Company for Freelance Community

Freelance Marketplaces are harnessing the Freelance talent and the Sector experienced tremendous growth during Severe Financial Crisis when companies were downsizing their employee base . But the Free-

lance Community was observed to be growing at Solid Speed opening doors to the Global world . With increasing number of Internet Users ,Mobile phone Users and 3G and 4G Customers , The Freelance Market has become the billion Dollar business since outsourcing has been termed as cost effective and fast way to get your Projects and tasks done without going into messy long time commitment with the Customers .

Considering the demand for new Freelance Marketplaces owing bursting and constantly growing Population , Workhoppers.com has been launched by Two Business Graduates and Entrepreneurs Vera and Linda as Startup Project with the plan to tap the local freelancers and get the tasks done.

Registration and Membership : Registration for becoming the mem-ber of Workhoppers for both Sellers and Buyers is Free .The Registration process is very easy and within few minutes ,you are on the go to use the powerful platform of Workhoppers . The Workhoppers won't waste your time in filling long forms of data but the basic information is required for registration .

Projects Categories : Workhoppers offers several freelance categories for Sellers and Buyers which include but limited to Business Consulting and Finance , IT and Software Projects (both customization and De-velopment) ,Designing and Media , Sales and Marketing , Writing and Translation , Administration ,Customer Support and other similar cate-gories . WorkHoppers has amazing platform to connect both Freelancers with Buyers . The Sellers can register for these categories where as Buyers can post their Projects in these related categories to get their tasks done

Payments : Workhoppers offer the secure and fats payment service . The buyers can pay the Sellers using Paypal and Standard Credit Cards such as MasterCard ,Visa , American Express where as the Sellers or Free-lancers get paid through using Paypal and Wire Transfer . The Payment System is also simple and Effective .

Pros: Workhoppers is free for both Freelancers or Sellers and For The buyers or Project Posters . Workhoppers has very sleek website which is great sign of Success as Growing Local Freelance marketplace. The num-ber of Freelance People is increasing day by day . The Volume of Freelance projects on Workhoppers is also growing with passage of time .

Cons: Workhoppers has limited payment system for Free-lancers or Sellers such as Paypal and Wire transfer . They should have Skrill , Pay-oneer for taping more Freelancers where Paypal is not supported . The Payment other FAQs must be updated with clear explantion of terms . The Time tracker for Hourly projects will add great value to Workhop-pers.com .

Future Growth Trends and Analysis : With growing number of Sell-ers and Buyers locally as well as Internationally , it is estimated that if Workhoppers sustained same rate of Growth and progress , It will soon stand in the list of Market Leaders such as Freelancer.com , Fiverr.com ,Elance and Odesk.com now Upwork.com provided that they followed the recommendation

5

New Freelance Marketplace :Linkedin Profinder

In 2016 the new Freelance Marketplaces are making their way to open the vistas of Opportunities . Thousands of freelancers across the globe use the Freelance Marketplaces to explore opportunities and connect

with buyers to get their tasks or Projects done online with perfection. The Internet has played vital role in bridging gap between countries

and providing the platform to connect Providers with Buyers and en-gaging in Business activity giving opportunity to home-based Moms and workers to earn their living by sitting in their homes .Some Freelancers are earning even more than those who are working in 9 to 5 Jobs in the offices . The new life style is being called the Dotcom Style which has en-gulfed almost every nation in the world and provided powerful platform to youth lancers .

Today , we are going to review the New Freelance Marketplace es-tablished by the Social Media giant Linkedin namely Profinder. Though Currently available to US based Freelancers yet it has clear plans to ex-pand this Powerful Freelance Marketplace to rest of World .

Establishment and The Founders : Linkedin.com/Profinder was established in year 2016 by Social Media giants Linkedin.com which offers jobs in all categories to Freelancers currently limited to US based Free-lancers soon to be expanded to other countries of the world .

Registration and Membership : The registration at is free at linkedin Profinder the Freelancers can apply for jobs and Buyers can post their Projects free . Both the Freelancers can easily sign in with their Linkedin Profile and engage in freelancing business instantly . No need of typing any thing or any information

Freelance Categories for Freelancers and Buyers : Profinder has sev-eral categories where the Freelance Professional ap-ply for freelance Pro-jcets and the buyers can post jobs for Freelancers to be done on Powerful and secure platforrm . The Categories include : Graphics Design ,Online Marketing ,Writing &Translation ,Video Animation , Programming , Advertising ,Business and several other related categories for the most de-manded Skills .

Payments :The Freelancers are paid by the buyers through Credit Card or Paypal where as the Freelancers Bank transfer or PayPal . The payment gateway is secure and requires fewer details and your are done .

NEW FREELANCE MARKETPLACE : LINKEDIN PROFINDER

UPside : Profinder is the biggest freelance marketplace to sell your skills and earn money. The Best thing is that you need not to register your details just sign using your linkedin Username and Password or even used Facebook account to Start working . Profinder is the great platform to connect freelancers with buyers and get the task done online .

DownSide : Profinder is great freelance marketplace which offers multiple categories to post projects and apply for free-lance jobs. Profinder lacks the feature of applying as pro from other countries except United States , Canada etc . Besides ,project posting by buyers , Profinder must add Skrill , wire Transfer and payoneer as payment solution as in some countries PayPal is not supported .

Future Trends and Analysis : With the rapid growth of the Web's one of the best Freelance Marketplaces ,it is estimated that if Profinder main-tained same growth , then it will surpass all the Competitors provided that it incorporates the suggestion made in this review in downside sec-tion . Linkedin.com/proFinder can succeed further if it expands its pow-erful Marketplace to Rest of the world .

6

Fiverr.com The Freelance Marketplace for Professional Gigs

Freelance Marketplaces are changing with the passage of time and every day, the new trends ,tools and frontiers are opening their doors of opportunities for the E-Workers for E-Business and earning E-currency . Millions of Freelance across the globe use the Freelance Marketplaces to explore the Freelance job stream and buyers around the world outsource the services of Professionals and get their tasks or Projects done online .

Thanks to Internet , which has transformed the in house Workers into Virtual Workers or home based workers and provided them with the Opportunity to earn their living .Some Freelancers are earning even more than those who are working in 9 to 5 Jobs in the offices . The new life style is being called the Dotcom Style which has engulfed almost every nation in the world .

We have a series of review the Freelance Startup Marketplaces and today we are going to review the growing Freelance

FIVERR.COM THE FREELANCE MARKETPLACE FOR PROFESSIONAL GIGS

Marketplace for Freelancers to sell their Skill based Gigs ranging between $5 to $50 Dollars or can be called the Micro Jobs . Fiverr.com is the world's biggest freelance Marketplace for Selling the GIgs and earn Money . Freelance Professionals post their Gigs and prospective buyers looking for the skills or service buy the gigs posted on fiverr.com .

Establishment and The Founders : Fiverr.com was established in year 2009 . The site was founded by by Micha Kaufman and Shai Wininge. The Fiverr.com offers the space to the Freelancers for selling their gigs starting from the as less as $5 . The Freelancers Sell their Gigs and buyers can easily buy the gigs posted by the freelance Professionals hailing from various countries of the world .

Registration and Membership : The registration at Fiverr.com is free fro the Freelancers to post their professional Gigs in the areas where they have great expertise provided that the Gigs adhere to Policy of Fiverr.com to maintain the Quality . The buyers can register for free and purchase the gigs at economical rates . They can get their tasks complet-ed through Fiverr.com .

Gigs Categories for Freelancers and Buyers : Fiverr.com has several categories to where the Freelance Professional post their categories easily and Sell them on the Powerful platform . There are about over one mil-lion Gigs posted on Fiverr.com by the Professional Freelancers . The Categories include : Graphics & Design ,Online Marketing ,Writing & Translation ,Video & Animation ,Music & Audio, Programming & Tech, Advertising ,Business and several other related categories for the most demanded Skills.

Payments for the Gigs Sell and Purchase : The Freelancers are paid by the buyers through Credit Card or Paypal where as the Freelance Gig Posters are paid by Fiverr.com by Bank transfer or PayPal. The pay-ment gateway is secure and requires fewer details and your are done. Fiverr.com has been growing rapidly due to its innovative Business Mod-el of Freelance Gigs or Micro Jobs as all other Similar Freelance website use Project posting and bidding format.

UPside : Fiverr.com is the biggest freelance marketplace to sell your skills based Gigs at small amount of money so that your gigs can be pur-chased easily and quickly and you keep on receiving the money stream without applying for each new project posted on the marketplace. Fiverr.com is providing you the service where the the employers contact the sellers and buy the gigs easily.

DownSide : Fiverr.com is the great freelance marketplace which of-fers the Professional Gigs posted by Freelancers . Fiverr.com lacks the fea-ture of posting Projects unlike Peopleperhour.com where Freelancers can post gigs as well as apply or bid for the projects which maximises the chances of freelancers to win the projects and earn extra money . Besides ,project posting by buyers , Fiverr.com must add Skrill , wire Transfer and payoneer as payment solution as in some countries PayPal is not supported .

Future Trends and Anslysis : With the rapid growth of the Web's one of the best Freelance Marketplaces ,it is estimated that if Fiverr.com maintained same growth , then it will surpass all the Competitors pro-vided that it incorporates the suggestion made in this review in downside section . Fiverr.com can further its growth by introduction ,project post-ing features and increasing the payment options so that freelancers can easily get paid .

7

Workana.com : The New Freelance Startup Marketplace

Freelancing has been the million dollar business and there have been countless Freelance marketplaces connecting Freelancers with the Buyers who are looking for verified and professional Writers ,Bloggers , Web Designers and programmers ,Sales and Marketing Professionals , Internet Marketers , Search Engine Optimization Specialist , Payroll and Human Resource Specialist , Translators ,Copywriters and several similar professions which could be done remotely . Thanks to Internet which has made Possible for the buyers and Providers to connect and gets the tasks done remotely .

There are Several Freelance Marketplaces helping the Buyers to get their projects completed remotely with great satisfaction but Workana is yet another addition to these Freelance Marketplaces . Established in 2013 , Workana has got the simple navigation and regular database of freelance projects along with best freelance professionals .

WORKANA.COM : THE NEW FREELANCE STARTUP MARKETPLACE

Membership/Registration: Registration for both the Freelancers and Buyers is free . Freelancers and buyers can easily register themselves . With simple steps you will be using the powerful Freelance marketplace of Workana.com .Workana also provides the facility of registration through facebook.com account which will obviously save your time . Workana.com also provides premium membership plans for freelancers so that the freelancer may apply for more jobs .

Projects and Time Tracker : There are several categories for the buy-ers to their projects and freelancers ti apply for the freelance jobs such as Writing and Translation ,Programming , Web Design and Development ,sales and Marketing , Admin Support ,engineering and Manufacturing ,Finance and Management and the Legal related projects . The buyers can monitor the freelancers through workana Time Report which pro-vides the time spent on the project .

Payments & Withdrawals : The buyers can pay for the completed projects through credit cards ,Paypal , DineroMail and Astro Pay . The payments from Workana are safe and secure . The Freelancers get paid through payoneer Prepaid Debit card and paypal .The Freelancers are paid twice a month by wokana .

Upside : Workana is the best addition to Freelance Marketplaces and may be competing with Odesk and Freelancer.com considered the Mar-ket Leaders in Freelance Marketplaces since Freelancer.com has eaten up its competitors through acquiring them with huge amount of money . Workana has the best platform may attract more Freelancers around the globe .

Downside : The Workana has got simple design which should be improved to attract more Freelancers and buyers to post their projects to get done by the Professionals . The Payments methods may be increased by adding Skrill and Direct bank payments . Workana should add its Contact address and Location so that Workana authority and reputation may be increased since hiding the Mailing Address or office Business will maker it doubtful .

Future Trends Analysis & Estimate : The Way Workana has devel-oped and attracted Hundreds of Freelancers and Buyers globally , it could b estimated that Workana will soon be competing with Freelance Marketplace Leaders such as Odesk.com , Elance.com and Free-lancer.com provided that they improved their features as mentioned in Downside .

8

Seven Secrets of becoming a Successful Freelancer

Freelancers are heading towards becoming the regular jobholders since the modern trends are evident that the Employers are only hiring those Freelancers who deliver the work on time and remain in constant contact with employers to be in their good books. Many newbies don't know the basics of freelance Marketplaces therefore ,lose the Pro-jects due to minor mistakes which cost them very heavy to sustain their Project base.

The Following Seven secrets will help the Freelancers specially the Newbies to get grip or hold on the Freelance jobs or projects. The secrets are very important and considered the key to be the most successful Free-lancer over the Internet since there are hundreds of Freelancers in vari-ous categories, but the only few freelancer pocket the perks of freelance Marketplaces while others manage to earn few bucks as part timers to buy some goods and coffee.

1. Be Specific :

Being specific to the category you feel that you are very good at and don't try meddle with multiple categories for which you have minimum or zero skills to to do the jobs . If you apply for such projects which are relevant to your experience and expertise , you may not succeed despite putting heart and soul in the project but the Employer is king , he may not accept the output you have delivered through limited Skills but the employer or buyer requires professionalism and the work should satisfy him as per requirement.If you boast about yourself that you can do multiple things and when project was awarded to you and you delivered the Poorest standard , you are likely to get bad marks from the buyers in form of Feedback . As some prospective buyers consider the feedback of any freelancer a key to hire the freelancers from global free-lancers pool .

2. Work Hours and Deadline:

The most important point for the Freelancers to specify their working hour or weekly limit of maximum hours a freelancer could do any job . Since some Buyers may require you to work full time as Virtual Assistants (VA's) or Project Managers as they remain busy in doing multiple things at a time . In such conditions , Be specific and clarify in the Cover letter or application if required by the buyer . You need not ignore the instructions of buyer in job posting as some buyers may decline the applications, then and there ,if you ignored the instructions in cover letter or bidding statement . Deadline is the point when you have to deliver the finessed and proofread document to the buyer on time , if you miss deadline , ultimately , you will miss the train bound to your destination so be on time and deliver one time to be a successful Freelancer.In order deliver on time , start early and finish be-fore the deadline as it will create positive points than delivering the date .

3.Be Honest and Polite :

Being honest and Polite is the key for successful Freelancer since if your brief the buyer about your skills in hon-est way , then there are 100% chances that you may win the bid or be awarded the Project right away . The Politeness while communication gives the impression that you are a polished experienced and Professional freelancers and ready to work for the project . The Polite request to offer some test service will further establish you as successful freelancers and you will experiencing tons of Projects coming in your way .

4. Communicate Regularly and Update the status :

here are several buyers and each buyer gets the Projects done in their own way some re-main close contact with Freelancer to pass the instruction for the service requested , where as ,some may contact you on the last day of the project for the output . But in both cases updating the buyers on daily basis is very necessary to be able to win their trust and inclination to hire you for longer period .You must check the dashboard regularly to insure that you may have received any instruction or message from the buyer . Answer Immediately as being non-responsive freelancer creates the impression or idea that you are not interested or irresponsible Freelancer and may receive negative feedback for communication skills . Update the buyer about the project document ,if .required ,share the part you have completed for buyers Review.

5. Asking Questions for Clearance :

If you have understood the requirements of the buyer then it is very nice . If you feel that you have some confusion about the project document or having any missing information then you must ask from the buyer to get it cleared before starting to work on the Assignment. As as many questions as you could related to the project to get it cleared so that you may not have to revise it when your buyer say that he wanted some points in it and you have not incorporated the same in the document . So , In order to escape such inconveniences, you must insure that you have understood the details of the pro-ject and you are ready to start work the project . Such gestures produce awesome results and the output is mostly accepted by the buyers if it was created by keeping the instruction in mind

6.Proof Read for Errors , Bugs or Typo mistakes:

Before sending the documents to the buyer first insure that the document you have created in free from typo , grammar or infatuation or Capitalization mistakes If you are freelance designers or Programmers , you must check it live whether the document has any programming errors or bugs so that you may fix them all and deliver an error free and perfect document to the buyers . sometimes , you buyer gets annoyed to receive the document which may be full of grammatical and TYPO errors and may not accept the document and may not pay you for the services delivered since your product was not proofread and spellings checked or bugs fixed .

7.Profile update and Portfolio and Custom Bid:

The most important and the most rewarding point is that you must have an updated portfolio on the Freelance Marketplaces and you must mention those skills which you believe that you are good at and can display the example of your past work or portfolio to make sure that you have mentioned only those skills which have already practiced and got positive feedback . In profile you may give career goals , length of experience and skills , guarantees for standard and quality of the work and links to your Blog ,content or web-sites you have worked for .

Don't lie or boast too much ,just presenting you skills to the buyers ,would be enough. Always create custom cover letters or custom bid statement keeping in mind the instruction of the job post . If you ignore the instruction and use recycled or copy pasted cover letter for all jobs , you will receive limited responses from the prospective buyers . In order to get positive responses , always create custom bid or proposal for the every new job you apply . Don't use the same proposal or cover letter for multiple jobs .

So, freelancers , we are summing this discussion up with following advice that if you keep above points in your mind to enter the lucrative and boundless world of Freelance , I am certain that you will certainly rule the roost in freelance World and I am afraid that seeing the great success , you may give up you day job . but don't make such blunder until you find enough resources and establish yourself as Professional free-lancer over the Internet .

9

Freelancing in 2018: Prospects and Opportunites

Freelancing or in other words working from home has been the great way of earning and making a way out of recession and financial crisis . The freelancers of countries with low opportunities for the labour have been able to find lucrative freelance projects working remotely from their homes .

The Trend in freelance has also attracted the full time employees and many of the full timers have become part timers and got rid of boss's angry looks and earned a lot of bucks by interacting with buyers of world.The Freelance Marketplaces such as Freelancer.com , Guru.com and odesk has really made their mark and become the top marketplaces with the projects posted followed by peopleperhour.com and scriptlance .

The Freelance Companies have connected millions of the freelancers with the prospective buyers who were Huntington fro qualified freelancers for getting their projects done . The projects with hourly payment were attractive than fixed freelance projects though fixed projects were more budgeted than those of hourly based projects.

The freelance Companies are eying more opportunities and more categories to attract more freelancers and Buyers by providing them powerful tools , secured and guaranteed payments solutions and many technological advances so that working online could be made reliable , cost effective and fun . The Freelancers choose their own time , own projects and own hours of working , this is,perhaps,the best and exciting way of working and earning at the same time .

The 2018 will prove very positive and fruitful for both freelancers and Buyers and freelance marketplaces but the recent draconian laws of axing the powerful medium of internet may affect its business opportunities since internet Marketers call internet the eyes and ears of information and possibilities .

The Freelance giants such as Freelancer.com, guru.com and upwork.com continue to rule the roost and succeeded in putting their grip on the lucrative business and have been updating their services and tools for both freelancers and buyers to excel each other in the fields of multi-national outsourcing . At the same time , there are many Freelance star-tups making their way and enter the competitive markets as new free-lance marketplaces . Hope 2018 will bring joys on the happy faces of free-lancers around the world with the message of peace and opening vistas of new opportunists and possibilities in this powerful business model.

10

The Social Media Importance For Freelancers & Entrepreneurs

The Social Media has great power which keeps on providing oxygen for new ventures and startups which are in great need of Traffic .Traffic is the only way to monetize and expand your brand visibility . The more visitors your site gets , the more perks will go in your ac-

count . Since people always look for the great brands which change their life style manifold . But the proper marketing of the product will yield more revenue and increase both demand and supply provided that mar-ket opportunity analysis is conducted and market trends are analyzed through research at grassroots level.

The Social media has really changed the picture Marketing and people from workers to executive Businessmen could be targeted geographically to announce your service . The social Networks such as facebook, Twitter , linkedin , facebook and myspace have taken the marketing specially social Marketing to next level where people share their interests and business announcement as well as other related product information for the guidance of prospective buyers online . The Social Media has established a connection between the companies and the customers to keep in touch with them all the time .

The Social Media gives you the facility to update the customers and visitors regarding your products , new products , improvements or any news regarding the company for their information . The Startups can up-date the customers and others by tweeting the products and news or any blog post . Similarly the facebook and Myspace also give you the convenience to share the updates regarding your company , products enhance-ments , renovation , improvements and technology so that the customers could be facilitated at every stage .

The small business owners can really benefit from this powerful medium of marketing since social marketing has been more impressive as it reaches the roots of the people .The Social Media can also be very beneficial for Individuals to target their employers or get a chance to be found through social networks such as linkedin .

The linkedin is really powerful medium to showcase your CV or resume along with skills so that the prospective employers may find you according your service category through various search engines such as Google , Yahoo , Bing , Ask and other search engines .If your are a consultant and if the employers search for consultant then your profile will be displayed to the buyers then if your profile matches their requirement there are almost chances you get hired . Thus social media really helps companies as well as individuals to get great exposure at no cost at all

11

Posting A Project On Freelance Marketplaces

Today, freelancing is a lucrative way to do many different types of jobs. For example, many freelance writers specialize in different types of writing. They can be employed on a freelance basis by websites, magazines, and even businesses that otherwise would not have writing staff. Many freelance web designers also work as web developers, graphic artists or programmers. Companies that need help with web designer graphics can employ these types of freelancers as contractors. Those companies can do so without having to employ new staff on either a short-term or temporary basis.

There are many freelancers online, and the Internet has made acces-sible a burgeoning online freelance marketplace. In this type of market-place, freelancers can make money by applying for a variety of jobs, usu-ally short-term and contract. Usually, companies or individuals post a "help wanted" ad on the freelance marketplace, or middlemen who post these types of jobs for others can also post these types of jobs. Once jobs have been posted, freelancers apply for them; in some cases, they bid for them instead of receiving a fixed price. Those with the best bid and/or the best experience usually get the job, whereupon that listing is closed. Those sites posting the jobs get a cut of the payment.

Freelancers are told that they should be conservative in the freelance marketplace, but they are also told that they should sell themselves and their skills well and should never be shortchanged. It's also true, however, that those posting the jobs need to take care as well. Of course, these peo-ple want someone who's willing to do the job for a fair price, and who
will return work in good quality. Both freelancers and those posting the jobs can get scammed, so this is where both parties need to be careful.

It's wise on both ends, therefore, to take care and make sure that you as a freelancer or job poster don't get taken advantage of. If someone posts a resume that simply has too much experience for the price he or she has bid, for example, this might be something to tell you to be care-ful. You should also take careful note of profiles that don't have much in the way of content.

Freelancers need to take care as well. Beware of clients or posters who promise extremely minimal pay for the job they want done. Make sure you state (and get) a fair price that will pay you well enough and fair-ly enough for a job well done—but then, you must deliver. If the poster does not choose you in this situation or wants you to go lower on your bid and you know it's not fair, walk away from the job or thank your lucky stars you didn't get it, since you don't want a job where you won't get fair pay.

As a new client or job poster, follow the freelance marketplace rules so that you know how to post projects properly. Each different market-place has its own rules and formatting, so that you should follow those. However, there are generally a few aspects of posting that are generally true almost every time. As a client or poster, you want to post your own name, a few details about your company, and a job description that's de-tailed enough for freelancers to understand what you want. Then, post a fair starting bid if applicable, or a fair price.

As a freelancer, there are also a few general rules that are usually true. You'll usually be asked to post a resume or other type of job/background experience. Oftentimes, you are also encouraged to post a portfolio so that prospective clients can see samples of your work. In some cases, you're also asked to post references.

Another point to remember is that even though bidding is meant to get a client or job poster the best price, the lowest bid will not always be the one you want. Remember that someone posting a very low bid may not have the experience you need, while someone with a higher but still reasonable bid knows what he or she is doing and is quoting a fair price. Therefore, choose based on the job experience of the freelancer and the bid rather than just the bid itself. You can also look at other bids on the website you choose to get a good idea of where you should place your starting bid.

Finally, make sure you get an account set up so that you can pay the freelancer quickly when the job is done. This will help ensure that your reputation as a client remains good so that freelancers will know that you are good to work for and will want to work for you; in turn, this will help ensure that your projects will be completed quickly, efficiently and to your specifications each time by experienced freelancers.

www.ingramcontent.com/pod-product-compliance
Lightning Source LLC
Chambersburg PA
CBHW031554210526
45464CB00003B/1295